Isomerism's Ingenious Interplay

Delving Into The World Of Isomers And Molecular Variations

Kiah Yaxunah

Contents

1 Introduction to Isomerism

Isomers—molecules sharing identical molecular formulas yet exhibiting distinct structures—form the bedrock of this exploration. Imagine building with LEGOs; you could use the same number of bricks, but arrange them to create vastly different structures—a car, a house, or a robot. Isomers are similar: the same atoms, different arrangements. This fundamental concept, this "same yet different," underlies the diverse and fascinating world of isomerism. We'll unpack this seemingly simple idea, revealing its complex and far-reaching implications.

Understanding the nuances of isomerism requires a systematic approach. We categorize isomers into two broad groups: structural isomers and stereoisomers. Structural isomers differ in the order their atoms connect. Think of it as rearranging the LEGO bricks to create different shapes altogether. On the other hand, stereoisomers have the same atom connectivity, but differ in the spatial arrangement of their atoms. It's like taking an identical LEGO structure and merely rotating or flipping certain parts. This seemingly subtle difference leads to surprisingly diverse properties.

Structural isomerism offers a fascinating glimpse into molecular diversity. Consider chain isomerism, where variations occur in the carbon skeleton's branching. Straight-chain alkanes possess linear structures, while branched-chain isomers show deviations from this linearity. This variation profoundly impacts the molecule's physical and chemical properties, influencing factors such as boiling points and reactivity. Positional isomerism involves shifts in the functional group's location on a carbon chain. A simple change in the functional group's position drastically alters reactivity and characteristics. Finally, functional group isomerism explores isomers with different functional groups altogether. This means the atoms are still the same, but their arrangement results in completely different functional groups with drastically varying chemical properties.

Stereoisomerism takes us into the realm of three-dimensional molecular architecture. Geometric isomerism arises from restricted rotation, often observed around double bonds or in cyclic structures. Cis and trans isomers, examples of geometric isomers, exhibit distinct properties due to the spatial positioning of their substituents. Consider the impact on melting and boiling points and their interaction with other molecules. Optical isomerism, a more intricate aspect of stereoisomerism, introduces the concept of chirality. Chirality describes the phenomenon of "handedness" in molecules, like left and right hands. These molecules, called enantiomers, are non-superimposable mirror images, interacting differently with polarized light and biological systems. The implications are vast, particularly in the field of pharmaceuticals, where the "wrong" isomer can have devastating consequences.

The implications of isomerism extend far beyond theoretical chemistry. Pharmaceutical applications demonstrate the importance of isomeric purity. Many drugs are chiral molecules, and only one enantiomer may possess the desired therapeutic effect, while the other may be inactive or even harmful. Understanding this is crucial in drug development and manufacturing. Furthermore, isomerism plays a significant role in natural products and biological systems. For instance, many naturally occurring compounds exist as isomers, and these forms often have distinct biological activities. The production of various polymers, essential in materials science and industry, is heavily influenced by isomerism and its relation to molecular properties. Knowing which isomer we're dealing with impacts everything from the polymer's strength and flexibility to its degradation rate. Indeed, understanding and controlling isomerism is vital for successful applications in these areas.

Delving deeper, we encounter more advanced concepts. Analyzing complex isomeric mixtures often requires sophisticated techniques, such as nuclear magnetic resonance (NMR) spectroscopy and mass spectrometry (MS). These techniques enable detailed structural characterization and identification of individual isomers within a mixture. Moreover, ongoing research expands our understanding of isomerism, exploring novel types of isomers and their potential applications. The continuous development of new analytical and computational tools pushes the boundaries of our knowledge, highlighting the dynamic and ever-evolving nature of this field. Understanding the fundamental principles laid out in this chapter empowers you to better grasp the vast and exciting world of isomerism, paving the way for further exploration of this compelling field.

1.1 Defining Isomers

Imagine you have a set of LEGO bricks. You have exactly the same number and type of each brick, but you can build many different things—a car, a house, a spaceship. Isomers are like that. They're molecules with the same number and type of atoms (the LEGO bricks), but the atoms are arranged differently, leading to different structures (the car, house, and spaceship). This fundamental difference in arrangement profoundly impacts their properties and behavior. We're not just talking about small changes; sometimes even the tiniest shift in an atom's position can dramatically alter a molecule's function, especially in biological systems. Understanding this principle is crucial for grasping the intricacies of chemistry.

This simple analogy, while helpful, doesn't capture the full complexity. Think about it—the arrangement of atoms extends beyond simply which atoms are connected to which. It encompasses the spatial orientation of atoms and groups of atoms in three-dimensional space. This is where it gets truly fascinating. The way atoms are oriented in space significantly influences how they interact with other molecules and their overall characteristics. For instance, certain spatial arrangements may give rise to entirely new chemical properties, leading to dramatically different responses in biological or chemical reactions. The effects cascade from the individual molecular level to larger scales like material properties and even macroscopic behavior.

Let's take a simple example: butane. It has the chemical formula CH. It's a hydrocarbon with four carbon atoms strung together in a chain and ten hydrogen atoms attached. However, there are two different ways to arrange these atoms, creating two isomers. One is a straight chain (n-butane), and the other is branched (isobutane). They both have the same formula, yet their boiling points differ—n-butane boils at a higher temperature due to its increased surface area and stronger intermolecular forces. This subtle difference in structure results in measurable differences in physical properties. This distinction underscores that isomerism isn't just a theoretical concept; it has real-world consequences.

The concept extends far beyond simple hydrocarbons. Imagine the complexity of large molecules like proteins or DNA. These contain thousands of atoms, and even slight alterations in their arrangements can lead to significant changes in their functionality. This difference explains why a subtle change in a protein's structure can cause a disease, or why two chemicals with the same formula can have dramatically different effects on the body – one might be a potent medicine, the other completely inert. Recognizing the difference between isomeric forms is therefore not just an academic exercise but fundamental to understanding the natural world and designing new materials.

Therefore, defining isomers isn't about memorizing a single definition. It's about internalizing the idea that molecules with the same molecular formula but different structures are fundamentally different entities. These differences manifest in varied ways—boiling point, melting point, reactivity, biological activity, and so on. The structural differences between isomers can be subtle but far-reaching in their consequences, significantly impacting various aspects of their behavior and chemical properties. Understanding this core principle is fundamental to comprehending the richness and complexity of the chemical world. Think of it as a puzzle: the same pieces, but infinitely many arrangements and therefore, infinitely many possibilities. The key to unraveling the puzzle lies in understanding the subtle but significant differences in how these pieces—the atoms—are arranged. Every variation, every shift in spatial arrangement, gives rise to a unique molecule with unique properties and behavior, waiting to be discovered and understood.

1.2 Types of Isomerism

Isomerism, at its core, is about molecular diversity. Imagine building with LEGOs – you could use the exact same bricks but create vastly different structures, from a simple house to a complex castle. Molecules do something similar. They share the same elemental composition, represented by their molecular formula, yet arrange themselves in different ways, leading to isomers, molecules with the same formula but distinct structures. This fundamental concept, though seemingly simple, opens doors to a world of structural and spatial variations.

Consider structural isomerism, where the arrangement of atoms differs. Picture three different ways to arrange the atoms of CH: a straight chain (butane), a branched chain (methylpropane) with a methyl group attached to the second carbon, and yet another branched-chain isomer. These isomers possess identical molecular formulas but different connectivities, impacting their physical properties like boiling points and melting points. Functional group isomerism takes this a step further; not only do atom connections change, but the type of functional group present shifts. For example, ethanol (an alcohol) and dimethyl ether (an ether) both have the formula CHO, but they exhibit different functionalities and, therefore, different chemical behaviors. They don't even react similarly. This subtle alteration leads to significant differences in reactivity.

Stereoisomerism paints a more intricate picture. Here, the atoms are connected identically, but their arrangement in three-dimensional space distinguishes them. Geometric isomerism provides a prime example. This kind of isomerism is a consequence of the restricted rotation around double bonds. Imagine two groups, X and Y, linked to carbons forming a C=C double bond; they can either be on the same side (cis) or on opposite sides (trans) of the double bond, creating two distinct geometric isomers with different properties. This restricted rotation has profound impacts on the properties of the isomers. For example, cis\item and trans- isomers of unsaturated fatty acids have different melting points, impacting their physical state at room temperature.

Optical isomerism dives even deeper into the three-dimensional world. Chirality, a key concept here, describes molecules that are non-superimposable on their mirror images, much like your left and right hands. These mirror-image isomers, called enantiomers, possess identical physical properties except for their interaction with plane-polarized light. One enantiomer rotates the light clockwise (dextrorotatory), and the other rotates it counterclockwise (levorotatory). This seemingly minor difference can have major implications in biological systems, as many enzymes show specific interactions with one enantiomer over another. Think about the thalidomide disaster: one enantiomer provided relief from morning sickness, while the other caused severe birth defects. This powerfully illustrates the importance of understanding optical isomerism.

Diastereomers, another type of stereoisomer, aren't mirror images of each other. They possess different spatial arrangements of groups around multiple chiral centers. They differ in physical properties as much as enantiomers differ, but they do not share the same optical activity properties. Meso compounds add another layer of complexity. Despite containing chiral centers, they possess an internal plane of symmetry, making them achiral and optically inactive.

Lastly, conformational isomerism, arising from the rotation around single bonds, introduces different spatial arrangements that are rapidly interconverting. Conformations like staggered and eclipsed forms in alkanes, or chair and boat forms in cyclohexanes, influence the molecule's stability and reactivity. The differences between these conformations are less significant than those in stereoisomers, but they still play a crucial role in molecular interactions and reactions. Understanding these nuances is critical for predicting molecular behavior. The understanding of different types of isomerism is vital for a multitude of applications, in various fields including medicine and drug design, understanding natural processes, and advancements in industrial processes and chemical manufacturing. The seemingly small difference between isomers can have massive consequences.

2 Structural Isomerism

Imagine building with LEGOs. You have the same number of bricks, but you can create vastly different structures – a car, a house, or a spaceship. That's essentially what structural isomerism is all about. It's the fascinating phenomenon where molecules share the same molecular formula, yet their atoms are arranged in different ways, leading to distinct properties and behaviors. This isn't just an abstract concept; it profoundly impacts how molecules interact with their environments, shaping everything from the scents of flowers to the effectiveness of medications.

Let's delve into the three main types of structural isomers. First, we have chain isomerism. Think of a simple hydrocarbon chain – you can arrange the carbon atoms in a straight line or branch them out in various configurations. These variations, while seemingly subtle, significantly alter the molecule's overall shape and, consequently, its properties like boiling point and reactivity. For instance, butane (CH) can exist as a straight-chain isomer or a branched-chain isomer, exhibiting different melting and boiling points, which impact their applications.

Next, picture yourself decorating a Christmas tree. Where you place the ornaments significantly changes its appearance. This is analogous to positional isomerism, where the position of a functional group or substituent on a carbon chain determines the isomer's identity. Consider 1-chloropropane and 2-chloropropane; both share the same molecular formula (CHCl), but the chlorine atom occupies a different position on the carbon chain, leading to distinct chemical reactions and physical characteristics, influencing their usage in various applications.

Functional group isomerism is a bit more dramatic – it's like swapping out entire sections of your LEGO creation. In this case, molecules possess different functional groups entirely, despite sharing the same overall molecular formula. For example, consider ethanol (CHOH), an alcohol, and dimethyl ether (CHOCH), an ether; they both have the formula CHO, but their vastly different functional groups result in dramatically different chemical and physical properties, making them suitable for entirely different applications. Ethanol is used in alcoholic beverages, while dimethyl ether is a widely used refrigerant.

Understanding structural isomerism isn't just about memorizing definitions; it's about gaining a deeper appreciation for the intricate relationship between molecular structure and properties. This understanding empowers you to predict the behavior of molecules based on their structure, a skill crucial across various scientific fields. From designing new drugs to developing more efficient materials, recognizing the subtle yet profound differences between structural isomers is a critical cornerstone of modern science and engineering. Consider the impact of even a single atom shift – a misplaced functional group might drastically alter a drug's effectiveness or a polymer's strength.

This knowledge is not confined to the laboratory; it touches our daily lives in countless ways. The scents and tastes of our food, the colors of our clothes, the effectiveness of medicines – these are all influenced by the intricate world of isomerism. The diversity arising from even slight rearrangements in molecular structure underpins the complexity and beauty of the chemical world surrounding us. The seemingly simple act of rearranging atoms unlocks a universe of possibilities, a testament to the ingenious interplay of isomerism in our world.

Think about the various applications of isomerism. Different isomers of the same compound often have dramatically different uses. Consider the case of limonene. Its isomers, R-(+)-limonene and S-(-)-limonene, are both present in citrus peels, but one gives off a refreshing, citrusy aroma, while the other has a more pungent, turpentine-like odor. This difference, arising from a simple spatial arrangement of atoms, allows them to find unique applications in various industries, ranging from food flavorings to cleaning products.

The same principle applies to countless other substances. Different isomers of a drug molecule can interact differently with the body, resulting in varying degrees of efficacy or toxicity. This is why understanding isomerism is particularly crucial in the pharmaceutical industry, as the development and safe use of drugs critically depend on identifying and characterizing the different isomers of a drug compound. Moreover, the subtle differences in the structural arrangement of polymer chains lead to variations in their physical properties, thus influencing their suitability for distinct applications, ranging from packaging materials to high-performance composites. Isomerism significantly influences the world around us.

In conclusion, mastering the concept of structural isomerism opens doors to deeper understanding in chemistry and beyond. It's not merely an abstract academic exercise, but a key to unlocking the secrets of molecular diversity and functionality. This knowledge empowers us to predict the behavior of molecules, design new materials, and develop innovative applications across various scientific disciplines, fundamentally influencing our world. By learning to see the subtle differences in structure, you unlock a world of exciting possibilities.

2.1 Chain Isomerism

Chain isomerism, a fascinating aspect of structural isomerism, involves molecules sharing the same molecular formula but differing in their carbon chain arrangement. Imagine building with LEGOs; you could use the same number of bricks but construct different shapes, like a long snake or a compact square. Similarly, molecules with identical atom counts can form distinct skeletal structures, leading to variations in their physical and chemical properties. This branching of the carbon skeleton drastically alters the molecule's overall shape and consequently, influences its boiling point, melting point, and reactivity. For instance, butane (CH) exists as two chain isomers: n-butane, with a straight chain, and isobutane, featuring a branched chain. This seemingly minor structural difference results in observable changes in their properties.
Unlike n-butane, isobutane is less susceptible to combustion, a key aspect often considered when selecting fuels for various applications. The compact structure of isobutane makes it less prone to extensive interactions, affecting its reactivity in different chemical environments. Understanding these distinctions between chain isomers is paramount in diverse fields, particularly in refining petroleum-based products. Refining processes often need to separate these isomers, as their differing properties influence fuel efficiency and combustion characteristics.

Consider the isomerism in the alkane family; as the number of carbon atoms increases, so does the number of possible chain isomers. Pentane (CH), for example, boasts three distinct chain isomers: n-pentane, isopentane, and neopentane. Each isomer exhibits unique properties. The increase in branching leads to a decrease in boiling points, due to reduced surface area for intermolecular forces to act upon. This variation in boiling points has major implications for separation techniques employed in the chemical industry. Different isomers might require different distillation temperatures, crucial for obtaining pure substances from complex mixtures.

The implications of chain isomerism extend beyond simply altering physical properties. Reactivity patterns also vary between different chain isomers. The accessibility of certain functional groups or carbon atoms is directly influenced by the shape of the molecule, thus affecting the rate and pathway of chemical reactions. This is especially significant in organic synthesis, where chemists need to carefully choose their reactants and reaction conditions to target specific isomers. Reactions are not always equally favorable for every isomer. Imagine, for example, a reaction requiring access to a specific carbon atom for substitution. A branched isomer might hinder this process compared to its linear counterpart.

Furthermore, the importance of chain isomerism extends into the realm of biological molecules, particularly lipids and fatty acids. The linear and branched structures of fatty acids influence their packing behavior in membranes and affect the fluidity of cell membranes. These structural differences affect how easily molecules can move across these cellular barriers. For example, saturated fatty acids, often possessing linear chains, tend to pack more tightly in membranes, whereas unsaturated ones, sometimes having more bends due to double bonds, lead to greater membrane fluidity. This impact on membrane properties underscores the crucial role of chain isomerism in biological systems.

The determination and distinction of these isomers rely heavily on sophisticated techniques such as nuclear magnetic resonance (NMR) spectroscopy, gas chromatography-mass spectrometry (GC-MS), and infrared (IR) spectroscopy. These methods allow chemists to decipher the intricate structures of these molecules and accurately identify different chain isomers. NMR spectroscopy, for instance, provides information on the connectivity and environment of the carbon atoms, allowing researchers to differentiate between linear and branched chains. GC-MS separates the isomers based on their boiling points and subsequently identifies them based on their mass-to-charge ratios, a crucial approach in analyzing complex mixtures.

The study of chain isomerism is not merely an academic exercise. It directly informs the design of fuels, the development of pharmaceuticals, and the understanding of biological systems. The distinct properties arising from variations in carbon chain arrangements have significant implications across various fields, highlighting the importance of this aspect of isomerism in both fundamental research and practical applications. From influencing fuel efficiency to shaping the function of cell membranes, the intricate world of chain isomerism continues to reveal surprising and vital information. As chemists continue to study and refine their understanding of these structural variations, we can anticipate further breakthroughs and innovative applications in chemistry and related disciplines. The possibilities are vast, and the implications far-reaching.

2.2 Positional Isomerism

Imagine you have building blocks, all the same shape and size. You could build a long, straight tower, or a shorter, wider structure. Both use the same number of blocks, the same type of blocks, but they look completely different. That's the essence of positional isomerism. We're talking about molecules, not towers, but the principle remains the same. In positional isomerism, the atoms are connected in the same order, but a functional group, or a substituent, occupies a different position within the molecule's structure. This seemingly small shift can significantly impact a molecule's properties, leading to variations in its reactivity, boiling point, melting point, and even its biological activity. Think of it like placing a sticker on a train – on one carriage, it's different from placing it on the engine, both use the same sticker.

Let's take butane as an example. Butane has the molecular formula CH. You can arrange these atoms in two distinct ways, forming n-butane and isobutane, both with the same atoms but a different arrangement. In n-butane, the four carbon atoms form a straight chain. Isobutane, however, features a branched chain, with three methyl groups attached to a central carbon atom. This difference in carbon chain arrangement is a type of structural isomerism, specifically, chain isomerism, which we discussed previously. In positional isomerism, the carbon skeleton remains the same, but the location of a functional group, for instance, a hydroxyl group (-OH) or a chlorine atom (Cl), changes, and that's why we're talking about it.

Consider the placement of a double bond within a molecule. For example, let's take but-1-ene and but-2-ene, both with the formula CH. Both isomers have a carbon-carbon double bond, but in but-1-ene, this double bond is at the very beginning of the carbon chain, at position one. In but-2-ene, the double bond sits between the second and third carbons. This seemingly minor change affects the molecule's reactivity. But-1-ene, for instance, is more reactive than but-2-ene due to the different electron density distribution around the double bond. Therefore, its chemical behavior varies. Understanding these differences is crucial in predicting their behavior in various chemical reactions and when used in industry or in the lab.

Positional isomerism isn't limited to simple hydrocarbons. It's prevalent across a vast spectrum of organic compounds, impacting molecules with more complex structures and a wider array of functional groups. For instance, consider the position of a nitro group (-NO) on a benzene ring. This can result in multiple isomers with different properties. You will find this position of a nitro group in common materials, like explosives or pharmaceuticals. 1-nitrobenzene and 2-nitrobenzene (also known as o-nitrobenzene) are just two of several possible isomers resulting from the placement of a single nitro group on a benzene ring. Such tiny differences in the position of this functional group significantly alter the molecule's reactivity and other chemical characteristics and their biological effects.

The impact of positional isomerism extends beyond simple physical and chemical properties. In the realm of biological systems, even subtle changes in the location of a functional group can dramatically affect how a molecule interacts with receptors or enzymes. This has profound consequences for pharmacology and drug design. For example, slight positional changes can transform an inert molecule into a potent drug, or vice versa, sometimes with life-threatening consequences. This emphasizes the crucial importance of understanding positional isomerism when dealing with biologically active compounds. Therefore, understanding positional isomerism is crucial in various areas such as medicinal chemistry and biochemistry.

Precisely identifying and distinguishing positional isomers necessitates the use of various analytical techniques. Spectroscopic methods, such as nuclear magnetic resonance (NMR) spectroscopy and infrared (IR) spectroscopy, are invaluable tools. NMR spectroscopy, in particular, provides detailed information about the environment of each atom within a molecule, enabling precise identification of the position of functional groups or substituents. Moreover, advanced separation techniques, such as gas chromatography (GC) and high-performance liquid chromatography (HPLC), can effectively separate and isolate individual positional isomers from a mixture, allowing for their independent characterization. These tools become essential for researchers and scientists. The ability to distinguish and characterize positional isomers is essential in various chemical disciplines, especially for quality control in industries using these substances.

To summarize, positional isomerism is a significant aspect of isomerism. It demonstrates that even minor changes in the position of functional groups within a molecule can have significant effects on its physical, chemical, and biological properties. Recognizing and understanding these subtle differences is critical for advancements in many scientific fields, including chemistry, biochemistry, and the pharmaceutical industry. Through the use of different analytical techniques, scientists can successfully identify and characterize different positional isomers, leading to a better understanding of their diverse behavior and properties. In conclusion, mastering positional isomerism is key to understanding the intricacies and diversity of the world of molecular structure and function.

2.3 Functional Group Isomerism

Functional group isomerism is a fascinating area of isomerism, where molecules share the same molecular formula but possess entirely different functional groups. This key difference profoundly impacts their chemical and physical properties, leading to vastly different behaviors. For instance, consider the molecular formula CHO. This formula can represent either ethanol, an alcohol with a hydroxyl (-OH) group, or dimethyl ether, an ether with an oxygen atom bonded to two methyl groups. Ethanol boils at 78°C and is a polar solvent, while dimethyl ether boils at -24°C and is significantly less polar. This dramatic difference highlights the power of functional group isomerism.

Understanding functional group isomerism requires a firm grasp of organic functional groups. Each functional group—alcohols, aldehydes, ketones, carboxylic acids, amines, and many others—imparts specific chemical reactivity to the molecule. Learning to recognize these groups within a molecular structure is crucial for predicting the compound's behavior. For example, the presence of a carboxylic acid group (-COOH) suggests acidic properties, while an amine group (-NH) points towards basic characteristics. The diversity in functional groups enables the creation of a vast array of molecules with widely varying properties from a single molecular formula.

Let's delve deeper into specific examples. Imagine two isomers with the formula CHO. One could be propanal, an aldehyde with the carbonyl group (C=O) at the end of the carbon chain. The other could be propanone, a ketone with the carbonyl group located in the middle. Aldehydes and ketones, while both containing a carbonyl group, differ in their reactivity due to the position of this group. Aldehydes are typically more easily oxidized than ketones. This simple yet crucial distinction showcases the significance of even subtle structural variations within functional group isomers.

Identifying functional group isomers might seem challenging initially, but a systematic approach simplifies the process. First, determine the molecular formula. Then, systematically explore the various possible functional groups compatible with that formula. Consider the number of carbon atoms, the possible arrangements of those atoms (chain isomerism), and the placement of the functional group (positional isomerism). Drawing out potential structures helps visualize the variations and solidify understanding. Software applications and online resources can also aid in generating and comparing different isomeric structures, greatly facilitating the learning process.

Moreover, the implications of functional group isomerism extend beyond simple organic chemistry. It's vital in fields like biochemistry, where subtle changes in functional groups dramatically affect the biological activity of molecules. For example, slight changes in the arrangement of functional groups in sugars can lead to different metabolic pathways and biological functions. Similarly, even minor differences in the structure of amino acids, the building blocks of proteins, are critical to a protein's three-dimensional structure and function. The study of functional group isomerism thus offers profound insights into the relationship between molecular structure and biological activity.

Think about the impact on drug design. Pharmaceutical companies extensively investigate functional group isomerism to create medications with enhanced efficacy and reduced side effects. Often, only one isomer within a set of functional group isomers demonstrates the desired therapeutic effect. Understanding these differences allows scientists to synthesize and utilize the active isomer, enhancing treatment efficiency and minimizing adverse reactions. By carefully considering functional group isomerism, pharmaceutical chemists can make a significant impact on patient health.

In conclusion, functional group isomerism is an important concept in chemistry with significant implications across diverse fields. It underscores the critical relationship between molecular structure and properties. By mastering the principles of recognizing and differentiating functional group isomers, you can unlock a deeper understanding of molecular behavior and open doors to numerous applications in chemistry and beyond. The ability to predict the chemical and physical properties of molecules based on their functional groups is a highly valuable skill that will serve you well in future studies and potential careers within the scientific realm. Remember, practice is key. The more examples you work through, the more confident and proficient you'll become in recognizing and understanding this vital aspect of isomerism.

3 Stereoisomerism

Imagine you have two LEGO models, both using the exact same bricks. However, one is a car, and the other is a house. They have the same components, but their arrangement creates completely different structures—that's the essence of isomerism. Stereoisomerism, a specific type of isomerism, takes this a step further. It's like having two LEGO models, identical in brick type and number, but arranged in ways that are mirror images of each other, or slightly different in their three-dimensional shapes, even though they have the same connections. These subtle spatial differences drastically change a molecule's properties.

Geometric isomerism, a key type of stereoisomerism, is like building two houses using the same LEGOs, but positioning the windows and doors differently. This difference results from restricted rotation. Imagine trying to rotate the walls of one house; it's fixed. In molecules, this restriction often arises from the presence of double bonds, where the pi bond prevents free rotation around the sigma bond, leading to distinct cis and trans isomers. Think of the double bond as a strong weld, holding parts in a specific orientation. Similarly, cyclic molecules experience restricted rotation because the atoms are locked in a ring structure. Understanding this restriction is fundamental to predicting a molecule's behavior.

Optical isomerism adds another layer of complexity. It's like building two identical LEGO houses, but creating mirror images that can't be superimposed. These are called enantiomers – they are not the same, even though their parts are identical. The key here lies in chirality – a molecule is chiral if it's non-superimposable on its mirror image, like your left and right hands. A carbon atom bonded to four different groups is a common source of chirality. These chiral molecules can interact with polarized light differently, rotating the plane of polarized light either to the left (levorotatory) or to the right (dextrorotatory). This ability to rotate polarized light is unique to chiral molecules and is used to distinguish enantiomers.

Consider the implications: even a tiny change in spatial arrangement can alter a molecule's biological activity. Enantiomers, despite their identical chemical composition, often exhibit dramatically different effects on living organisms. One enantiomer of a drug might be highly effective, while its mirror image could be inactive or even toxic. This is why the pharmaceutical industry invests heavily in creating single enantiomer drugs, thus enhancing efficacy and safety. It's a critical area of research, ensuring drug development targets the desired spatial configuration and avoids potentially harmful effects. This highlights the profound impact of stereoisomerism.

Understanding stereoisomers goes beyond simply recognizing their differences. Predicting the behavior and properties of these molecules requires delving into their specific arrangements. The cis and trans isomers of alkenes, for example, often have markedly different melting points and boiling points. Furthermore, their chemical reactivity can also vary significantly, influencing their applications in different areas. This means choosing the right stereoisomer for a specific task is crucial—much like selecting the appropriate tool for a job, you cannot simply use any tool for any job.

Imagine trying to design a new drug or polymer. Understanding stereoisomerism will help you predict the properties of your potential creation. It would be a critical consideration in drug design, materials science, and numerous other fields. The subtle yet profound differences created by these spatial arrangements highlight the importance of understanding stereoisomerism in a wide range of scientific disciplines. For instance, in polymer chemistry, the stereoisomerism of monomers significantly impacts the final polymer's properties like strength, flexibility, and crystallinity. This is because the way the monomers are arranged affects the overall structure and interactions within the polymer chain.

Beyond practical applications, the study of stereoisomerism helps us understand nature's remarkable complexity. Many naturally occurring molecules, such as sugars and amino acids, exist as chiral molecules with specific spatial arrangements. The chirality of these molecules is critical for their biological activity and interactions within living organisms. Thus, the study of stereoisomerism is essential for fields such as biochemistry, where the specific three-dimensional structure of a molecule often dictates its function in biological systems. Essentially, mastering this concept offers a deeper appreciation of the intricate world of molecules and their functions. The remarkable relationship between structure and function is central to the beauty and complexity of the chemical world. Mastering stereoisomerism opens doors to understanding this connection fully. Therefore, the significance of stereoisomerism extends far beyond the realm of pure chemistry; it provides a foundation for comprehending a world teeming with molecules that interact in complex and fascinating ways.

3.1 Geometric Isomerism

Imagine you're building with LEGOs. You have the same number of bricks, but you can build a tall, thin tower, a wide, short wall, or even a sprawling castle. That's similar to geometric isomerism. Geometric isomers are like those different LEGO creations—they share the same atoms and bonds, but their spatial arrangement differs, impacting their properties significantly. This difference arises from restricted rotation around a double bond or within a ring structure, creating distinct molecular shapes. You can't just twist one geometric isomer into the other, unlike some other types of isomers. It's a fixed arrangement.

Think of a simple alkene, like but-2-ene. It has a carbon-carbon double bond. Around this bond, rotation is restricted. This restriction leads to two geometric isomers: cis-but-2-ene and trans-but-2-ene. In cis-but-2-ene, the two methyl groups are on the same side of the double bond, causing a slight bending in the molecule's shape. This configuration results in stronger dipole-dipole interactions compared to its counterpart. In trans-but-2-ene, however, the methyl groups are on opposite sides, leading to a linear, or almost linear, molecule with weaker interactions. These subtle differences cause variations in their melting and boiling points, reactivity, and even their physical states at room temperature. Therefore, you can have two different materials having the same formula, one a liquid and the other a gas.

Cyclic compounds provide another fantastic illustration of geometric isomerism. Consider 1,2-dimethylcyclopropane. Because the carbon atoms in the ring are connected in a fixed way, the two methyl groups can either be on the same side of the ring (cis) or opposite (trans). Just like with alkenes, these spatial arrangements result in different physical and chemical properties. The cis isomer, for instance, might experience steric hindrance, a repulsive interaction between the methyl groups. This hindrance can influence its reactivity and stability, making it potentially less stable than its trans counterpart. Cis and trans isomers demonstrate how a seemingly small change in molecular structure can lead to drastically different properties.

The impact of geometric isomerism isn't limited to the laboratory; it extends to the natural world and industrial applications. In the realm of vision, retinal, a molecule crucial to sight, exists as a cis-trans isomer. Light absorption triggers a change from cis to trans, initiating a cascade of events that ultimately lead to signal transmission to the brain. This fundamental isomerization is essential for the perception of light. Similarly, in the synthesis of polymers and other important materials, controlling the geometric isomerism of the building blocks is vital for dictating the properties of the final product. For instance, the degree of cis or trans arrangement in polyisoprene significantly influences the physical properties of the resulting rubber, affecting its elasticity and strength.

Geometric isomers can be identified using various spectroscopic techniques. Nuclear Magnetic Resonance (NMR) spectroscopy, for example, can distinguish between cis and trans isomers by analyzing the chemical shifts and coupling constants of their protons. Infrared (IR) spectroscopy can also be useful, showing differences in vibrational frequencies between the isomers due to variations in their bond angles and dipole moments. Understanding these techniques is essential for characterizing and identifying geometric isomers in different settings. This detailed analysis plays a crucial role in several fields like medicine, where distinguishing between geometric isomers can be lifesaving, as their biological activities often differ

significantly. Some isomers may display therapeutic effects, while others may be toxic. Precise identification is therefore paramount.
Furthermore, the concept of geometric isomerism extends beyond simple alkenes and cyclic compounds. It's observed in various organic molecules with hindered rotation around specific bonds and in coordination complexes in inorganic chemistry. The principles discussed remain relevant, highlighting the pervasive nature of this type of isomerism in a wide range of chemical structures and phenomena. This versatility makes geometric isomerism a topic of great importance and interest across different subdisciplines in chemistry. The subtle yet significant differences between geometric isomers emphasize the intricate relationship between molecular structure and properties, showcasing the power and subtlety of isomerism.

3.2 Optical Isomerism

Imagine a pair of hands. They're mirror images, right? Identical in every way, yet you can't superimpose one perfectly onto the other. That's the essence of optical isomerism, a fascinating type of stereoisomerism where molecules exist as non-superimposable mirror images, called enantiomers. These aren't just abstract concepts; they have real-world implications, particularly in the field of pharmaceuticals.
Understanding optical isomerism requires grasping the concept of chirality. A chiral molecule possesses a chiral center, usually a carbon atom bonded to four different groups. This asymmetry is what allows for the existence of enantiomers. Think of it like a molecule with a 'handedness'—a left-hand version and a right-hand version, each distinct from the other. This 'handedness' is crucial, especially when the molecule interacts with other

chiral molecules, such as receptors in your body.

One of the most striking differences between enantiomers lies in their interaction with plane-polarized light. Plane-polarized light, unlike ordinary light, vibrates in only one plane. When this light passes through a solution of a single enantiomer, the plane of polarization rotates. One enantiomer will rotate the light clockwise (dextrorotatory, denoted as +), while its mirror image will rotate it counterclockwise (levorotatory, denoted as -). This property is called optical activity and provides a way to distinguish between enantiomers.

Now, let's consider what happens when you have an equal mixture of both enantiomers. This is called a racemic mixture. Intriguingly, a racemic mixture doesn't exhibit optical activity because the rotations caused by each enantiomer cancel each other out. This is like having equal numbers of left and right hands—the overall effect is no specific handedness. However, while optically inactive as a whole, racemic mixtures can exhibit different physical properties compared to pure enantiomers. Their melting points, for instance, may differ.

The biological activity of enantiomers can be remarkably different. Consider the infamous thalidomide tragedy. Thalidomide exists as two enantiomers, one with sedative properties and the other with teratogenic effects (causing birth defects). The drug initially marketed was a racemic mixture, leading to devastating consequences. This highlights the critical importance of considering the stereochemistry of drugs during development and testing.

Beyond pharmaceuticals, optical isomerism plays a significant role in various fields. In the food industry, enantiomers often have different tastes or smells. Consider the example of carvone, one enantiomer of which smells like spearmint while the other smells like caraway. This difference is due to the distinct interactions of the enantiomers with receptors in our noses. Similarly, many natural products, such as sugars and amino acids, exist predominantly as single enantiomers.

This specificity in biological systems is partly due to the chiral nature of enzymes, proteins that catalyze reactions in living organisms. Because enzymes are chiral, they typically interact more strongly with one enantiomer than with the other, exhibiting stereoselectivity. This stereoselectivity is crucial for many biological processes, such as the synthesis and breakdown of molecules within our bodies. This interaction is not simply a matter of chance; it is directly related to the precise three-dimensional structure of both the enzyme and the substrate.

Understanding optical isomerism is not just an academic exercise; it has profound implications across diverse disciplines. From drug design and development to the study of biological processes and even the flavors we experience in our food, this intriguing aspect of molecular structure has far-reaching consequences. Mastering this topic empowers you to view molecules with a new level of depth and appreciation for their complexity and the subtleties of their interactions with the world around them. As you continue your journey through the realm of isomerism, remember that the seemingly small differences in molecular structure can have vast and significant effects. The power of subtle variations in molecular architecture is truly remarkable. Recognizing and understanding these differences is key to advancing our understanding in many scientific fields.

4 Chirality and Enantiomers

Imagine a pair of hands. They're mirror images, right? Yet, you can't superimpose one perfectly onto the other. That's the essence of chirality. In chemistry, a chiral molecule is one that lacks an internal plane of symmetry, like your hands. This seemingly simple concept has profound implications for the properties and behavior of molecules, especially in biological systems. We're talking about a difference that's visually subtle but functionally significant, a difference that impacts everything from drug effectiveness to the way organisms function.

Enantiomers are pairs of these chiral molecules – they're non-superimposable mirror images of each other. Think of them as those left and right hands again. They possess identical physical properties, like melting point and boiling point, under normal conditions; however, their interaction with other chiral molecules is where the magic – or sometimes the mischief – happens. This selective interaction is crucial because biological systems are often chiral themselves, meaning they prefer to interact with one enantiomer over the other.

Consider the infamous thalidomide tragedy. This drug, used to alleviate morning sickness in pregnant women, existed as a mixture of two enantiomers. One enantiomer had the desired effect, while the other caused severe birth defects. This tragic example highlights the critical importance of understanding chirality in drug development and emphasizes the necessity for separating and purifying individual enantiomers for pharmaceutical use, ensuring the desired effects and preventing harmful side effects. The differences are incredibly subtle, yet the consequences can be disastrous.

Now let's delve into racemic mixtures. This term simply means a 50/50 mixture of both enantiomers. It's like having an equal number of left and right hands. While the overall mixture might seem to have balanced properties, the individual enantiomers retain their distinct characteristics, leading to a mixture of effects rather than a unified, predictable outcome. Understanding racemic mixtures is important because many chemical synthesis processes initially produce these mixtures; therefore, resolving them into their individual components—a process called chiral resolution—is often a crucial step in many applications.

But it gets even more intricate. There are other types of stereoisomers beyond enantiomers, including diastereomers. Unlike enantiomers, which are mirror images, diastereomers are stereoisomers that are not mirror images of each other. They arise from molecules with more than one chiral center. This leads to a fascinating array of possibilities and distinct properties. The differences between diastereomers can be more significant than those between enantiomers, influencing physical properties and reactivity in diverse ways. Separating diastereomers can be easier than separating enantiomers because their differences in properties are often more pronounced.

Finally, let's discuss meso compounds. These are molecules that possess chiral centers yet are achiral themselves—they have an internal plane of symmetry. It's like a molecule that cleverly cancels out its own chirality. Even though they contain chiral carbons, their overall symmetry negates the effects of chirality. These molecules often defy our initial intuitive understanding of chirality and require a careful analysis to identify their achiral nature. Their presence in a mixture necessitates further scrutiny of the stereochemical composition of a molecule's isomers. Their existence is a testament to the subtlety and complexity of the relationship between structure and chirality. Mastering the concepts of chirality, enantiomers, diastereomers, and meso compounds is critical for navigating the diverse world of isomerism and understanding its influence across various scientific disciplines.

4.1 Defining Chirality

Imagine you're building with LEGOs. You have the same number of bricks, but you can construct a car, a house, or a robot. That's similar to isomerism; molecules with the same formula but different structures. Chirality is a special kind of isomerism, focusing on a molecule's "handedness." It's a property of asymmetry, like your left and right hands. They look alike, but you can't perfectly superimpose them.

Think about a carbon atom bonded to four different groups. This carbon is called a chiral center. It's like having a central point with four unique branches extending in different directions. Because of the arrangement of these four groups, the molecule becomes non-superimposable on its mirror image. This creates two different forms, called enantiomers. They're mirror images, yet distinct, much like those left and right hands.

You might think that since they're mirror images, they would behave identically. That's not true! This is where the fascinating aspect of chirality comes in. Enantiomers often exhibit strikingly different properties when interacting with other chiral molecules. For example, consider how our bodies are built from chiral molecules. Proteins, DNA, and many other vital components are chiral. These chiral molecules often interact differently with each enantiomer of a drug, leading to varied effects. One enantiomer might be effective medicine while the other is inactive or even toxic.

To understand this difference, imagine trying to fit a left-handed glove onto your right hand. It won't work. Similarly, a receptor site in your body, which is often chiral, may perfectly match one enantiomer of a drug, allowing it to bind and exert its biological effect. The other enantiomer, however, may not fit and will have either no effect or possibly a negative interaction. This is crucial in pharmaceutical development, where chemists carefully synthesize and isolate the desired enantiomer for optimal efficacy and safety.

Let's explore how we can identify a chiral molecule. The presence of a chiral center, a carbon atom bonded to four different groups, is a key indicator. However, not all molecules with a chiral center are chiral. Some molecules possess internal symmetry that cancels out the chiral effect. These are called meso compounds. They have chiral centers but are achiral due to a plane of symmetry within the molecule, rendering them superimposable on their mirror images despite the chiral centers. They're like optical illusions, seeming asymmetrical but revealing a hidden symmetry.

Beyond the chiral carbon, we can examine the overall symmetry of the molecule. If a molecule has a plane of symmetry – imagine cutting it in half so that one half is a mirror image of the other – it will not be chiral. This symmetry renders the molecule superimposable on its mirror image, negating the property of handedness. Understanding molecular symmetry is essential for determining chirality; many molecules are more complex

and have several chiral centers.
Finally, remember, chirality is not merely an academic curiosity. It's a cornerstone of biological activity. Enzymes, the workhorses of biochemical processes, are usually chiral and interact selectively with chiral substrates. This selectivity is the basis for many biological processes, from metabolic pathways to drug action. Consequently, understanding chirality is crucial not only for organic chemists but also for biochemists, pharmacologists, and anyone interested in the intricacies of life at the molecular level. The concept of chirality, thus, provides a glimpse into the elegance and precision of nature's design. Mastering this concept unlocks a deeper understanding of molecular interactions and their profound implications.

4.2 Properties of Enantiomers

Enantiomers, as you've learned, are molecules that are non-superimposable mirror images of each other. This seemingly small difference leads to some surprisingly significant variations in their properties, impacting everything from their smell to their biological activity. Understanding these differences is key to grasping the importance of chirality in chemistry.
One of the most striking properties of enantiomers is their ability to rotate plane-polarized light. This is a phenomenon you can observe experimentally using a polarimeter. Each enantiomer rotates the plane of polarized light in an equal but opposite direction, a property designated as either dextrorotatory (+) or levorotatory (-). This optical activity is a direct consequence of their three-dimensional structure and a crucial characteristic for identification and characterization.
However, optical activity isn't the only difference; enantiomers can exhibit

distinct physical properties. While their melting points and boiling points are usually identical, their interactions with other chiral molecules differ significantly. Think of it like trying to fit a left hand into a right-handed glove; it simply won't work. Similarly, a specific enantiomer might interact strongly with a chiral receptor in your body, while its mirror image may have little to no effect, or even a completely opposite effect. This is crucial in pharmacology.

Consider the infamous thalidomide tragedy. Thalidomide exists as two enantiomers, one with sedative properties, and the other with teratogenic properties, causing severe birth defects. The initial drug formulation was a racemic mixture, containing both enantiomers, resulting in devastating consequences. This tragic event highlighted the critical importance of understanding and controlling the chirality of pharmaceuticals.

The differences extend beyond biological interactions. Enantiomers can also interact differently with other chiral molecules in chemical reactions. This selectivity is harnessed by chemists in asymmetric synthesis, where the goal is to create a single enantiomer rather than a racemic mixture. Enzymes, which are naturally chiral, often exhibit high levels of enantioselectivity. This means they can catalyze reactions preferentially with only one enantiomer, allowing for the synthesis of pure enantiomers for various applications.

Moreover, the separation of enantiomers, known as chiral resolution, presents a significant challenge. Because their physical properties are often very similar, separating them requires specialized techniques. These include chiral chromatography, where enantiomers are separated based on their differential interactions with a chiral stationary phase, or the formation of diastereomers with a chiral resolving agent, which then allows for

easier separation based on differences in their physical properties.
The interaction of enantiomers with polarized light is a key differentiating factor. A racemic mixture, a 50:50 mixture of both enantiomers, doesn't rotate plane-polarized light. The rotations of the individual enantiomers cancel each other out, resulting in zero net optical rotation. This is a crucial observation, demonstrating how the macroscopic properties of a mixture depend heavily on the presence and ratios of each enantiomer.
The implications of enantiomeric differences extend far beyond the laboratory. In nature, many biological molecules, like amino acids and sugars, predominantly exist as a single enantiomer. The reasons for this chiral bias are complex and still being actively investigated. But, the consequences are clear; life as we know it relies heavily on this selectivity. Consider the proteins in our bodies, the complex structures formed from specific sequences of amino acids. A change in the chirality of just a single amino acid could drastically alter the protein's structure and functionality, and that difference could have wide-reaching consequences.
Finally, remember, while enantiomers share the same connectivity, their different spatial arrangements make them distinct chemical entities with unique and sometimes drastically different properties. Understanding these differences is not simply an academic exercise but is crucial for various applications, from designing effective pharmaceuticals to understanding the intricacies of biological systems. Their study highlights the fascinating interplay of structure and function at the molecular level. The seemingly subtle difference between a mirror image and its original profoundly shapes the world around us.

4.3 Racemic Mixtures

Imagine you're baking a cake. You carefully measure out all the ingredients – flour, sugar, eggs, butter – following the recipe precisely. Now, imagine you have two bowls, both containing exactly the same ingredients in exactly the same quantities. But in one bowl, the ingredients are mixed thoroughly, creating a uniform batter. In the other, some ingredients clump together while others are isolated – a less than ideal cake batter. Both bowls have the same composition, the same molecular formula, if you will, but they are demonstrably different. This analogy perfectly encapsulates the concept of a racemic mixture.

A racemic mixture is like that second, uneven cake batter. It's a 50/50 mix of two enantiomers, those mirror-image isomers we discussed earlier. Think of them as left and right hands; they're identical in every way except for their spatial orientation. They are non-superimposable mirror images, just like your hands. You can't overlay them perfectly. These enantiomers, despite their identical chemical formulas, often display strikingly different properties. One might taste sweet, the other bitter; one might be a potent medicine, the other entirely ineffective, or even toxic.

This is because of their interaction with other chiral molecules. Remember, many biological molecules—enzymes, receptors, and so on—are themselves chiral. This means they exhibit a "handedness," and they interact differently with each enantiomer. It's like trying to fit a right-handed glove onto your left hand—it just doesn't work. Consequently, the biological activity of a drug, for example, might entirely depend on which enantiomer it contains. In a racemic mixture, both enantiomers are present, potentially leading to unpredictable and possibly harmful effects if only one is biologically active.

This unequal biological activity has immense implications in pharmacology. For instance, thalidomide, a drug once used to treat morning sickness, was marketed as a racemic mixture. While one enantiomer relieved morning sickness, the other caused severe birth defects. This tragic case highlights the critical importance of understanding and separating enantiomers, especially in drug development. Before a drug is approved for use, extensive testing is performed to assess the efficacy and safety of each individual enantiomer, ensuring a thorough understanding of their potential impacts. It is not enough to simply test the racemic mixture.

The separation of enantiomers, known as chiral resolution, is a significant challenge in organic chemistry. Several techniques exist, including chiral chromatography, where enantiomers are separated based on their different interactions with a chiral stationary phase. Another method utilizes diastereomers. By reacting a racemic mixture with a chiral reagent, you form a mixture of diastereomers, which have different physical properties and can be easily separated using traditional techniques like crystallization or distillation. Once separated, the individual enantiomers can then be purified and their properties further characterized.

Think about it – the consequences of dealing with a racemic mixture without understanding its composition can be drastic. Therefore, identifying and separating the individual enantiomers within a racemic mixture is not just an academic exercise; it's an essential step for developing safe and effective drugs, synthesizing new materials, and even understanding fundamental biological processes. This careful approach leads to a more precise understanding of the molecule's true nature and function, preventing potentially harmful outcomes. The importance of this separation can't be overstated; it's a critical step in ensuring the safe and effective use of chiral molecules across many sectors.

Racemic mixtures represent a fascinating aspect of isomerism. The seemingly subtle difference between mirror-image isomers can lead to profound variations in their properties, highlighting the importance of understanding the three-dimensional structure of molecules. As you delve deeper into the world of isomerism, remember the crucial role of enantiomers and the potential pitfalls—and profound applications—of racemic mixtures. Their study provides a clear illustration of how minute structural changes can lead to entirely different outcomes, impacting fields from medicine to materials science. Mastering the ability to identify and resolve racemic mixtures is a testament to your growing understanding of the intricate world of molecular structures and their interactions.

4.4 Diastereomers

Diastereomers are stereoisomers that are not mirror images of each other. Unlike enantiomers, which are non-superimposable mirror images and possess identical physical properties except for their interaction with plane-polarized light, diastereomers exhibit distinct physical and chemical properties. This difference arises from their unique spatial arrangements, leading to variations in their interactions with other molecules. Think of it like comparing a right-handed glove to a left-handed glove (enantiomers) versus comparing a right-handed glove to a right-handed mitten (diastereomers)—both are right-handed, but they're clearly different.

Consider a molecule with two chiral centers. Each chiral center can exist in two configurations, R or S. This leads to a total of four possible stereoisomers. Two of these stereoisomers will be enantiomers (mirror images), while the other two pairs will be diastereomers. They are not mirror images, and this fundamental difference profoundly impacts their properties. For instance, one diastereomer might be more soluble in water than the other, or they might react at different rates with the same reagent. Such differences are crucial in various applications.

The existence of diastereomers significantly expands the diversity of isomeric forms and the complexity of chemical systems. Their distinct characteristics allow for precise manipulation and control over chemical reactions and molecular interactions. Imagine the possibilities in designing pharmaceuticals: the subtle difference between diastereomers might translate to a dramatic shift in biological activity, with one isomer being a potent drug while another is inert, or even toxic. This highlights the importance of understanding diastereomerism in drug development. Careful separation and characterization of diastereomers are essential steps in drug design and synthesis. The field of medicinal chemistry is full of examples where a single diastereomer demonstrates therapeutic value, while the others do not.

Diastereomers are not limited to organic molecules; they can also be observed in inorganic complexes and organometallic compounds, demonstrating the widespread nature of this phenomenon. Their presence influences the physical properties of materials, potentially leading to novel applications in various industries. For instance, their unique optical properties and structural features might contribute to the development of new materials for optoelectronics or catalysts with improved selectivity. In materials science, subtle variations in diastereomeric configurations can significantly affect the properties of polymers and crystals. This broadens the scope of diastereomerism far beyond organic chemistry.

Determining the configuration of diastereomers often involves sophisticated techniques such as X-ray crystallography and nuclear magnetic resonance (NMR) spectroscopy. These methods provide detailed structural information, allowing scientists to differentiate between diastereomers and fully characterize their configurations. Advanced spectroscopic techniques, coupled with computational modeling, are crucial tools in understanding the complex interactions that arise from diastereomeric relationships. This interplay between experimental data and theoretical predictions is paramount in developing a comprehensive understanding of diastereomerism's impact across various chemical disciplines.

The study of diastereomers doesn't merely provide an academic exercise; its impact extends to practical applications. Separating diastereomers is often crucial for various industries, including pharmaceuticals and food production. Think about the stringent purity requirements in pharmaceuticals – only the desired diastereomer with the specific biological activity can be tolerated; the others must be removed. Similar considerations apply to other sectors where the characteristics of specific diastereomers determine the final product's quality and performance. These are all examples where a fundamental understanding of diastereomerism is not merely an academic exercise; rather, it is a critical prerequisite for ensuring the effectiveness and safety of various products.

In conclusion, diastereomers represent a significant aspect of stereoisomerism, offering a deeper understanding of the relationship between molecular structure and properties. Their unique characteristics provide valuable opportunities for advancements in various fields, highlighting the importance of this topic in organic chemistry, materials science, and medicinal chemistry. The ability to understand, identify, and manipulate diastereomers opens new avenues for innovation, leading to the development of new materials and pharmaceuticals with precisely tailored properties. The journey into the world of diastereomers is a journey into the intricate details that ultimately dictate the behavior and application of countless molecules. This intricate understanding paves the way for future breakthroughs in diverse fields.

4.5 Meso Compounds

Meso compounds are a fascinating subset of stereoisomers, specifically diastereomers, possessing internal symmetry. Imagine a molecule that's chiral, meaning it has a non-superimposable mirror image, but somehow manages to be achiral overall. That's the essence of a meso compound. This internal symmetry cancels out the chiral effects of individual parts, leading to an overall lack of optical activity; they don't rotate plane-polarized light. Think of it like having two opposing forces perfectly balanced, resulting in no net movement.

This internal symmetry usually arises from a molecule containing multiple chiral centers, but with a plane of symmetry dividing the molecule into two identical halves. Each half is a mirror image of the other, perfectly reflecting across that central plane. This mirror image relationship within the single molecule is what negates the optical activity associated with typical chiral molecules. This is different from a mixture of enantiomers (a racemic mixture), which shows no optical activity because the rotational effects of the two enantiomers cancel each other out. A meso compound, however, is a single molecule with inherent symmetry leading to zero optical rotation.

Let's consider a simple example – tartaric acid. It has two chiral centers, meaning two carbon atoms each bonded to four different groups. While most molecules with two chiral centers would have four stereoisomers (two pairs of enantiomers), tartaric acid only has three: two enantiomers and one meso compound. The meso form possesses a plane of symmetry running through the molecule, splitting it into two identical, mirror-image halves. This inherent symmetry means the molecule cannot rotate plane-polarized light, a defining characteristic of achiral compounds. This demonstrates how the presence of multiple chiral centers doesn't automatically guarantee optical activity. The crucial factor is the presence or absence of a plane of symmetry, and this can sometimes be surprisingly subtle to visualize. One needs to carefully look at the 3D structure to identify if there's any such symmetry.

Understanding meso compounds is crucial for grasping the nuances of stereochemistry. It highlights that optical activity isn't solely dependent on the number of chiral centers but significantly on the overall symmetry of the molecule. Many organic chemistry students initially struggle with visualizing molecular symmetry, but practice and the use of molecular modeling tools can enhance this ability considerably. Different models and representations, whether ball-and-stick models or software simulations, are helpful in comprehending the three-dimensional arrangements responsible for meso compound characteristics.

Unlike enantiomers, which interact differently with other chiral molecules, meso compounds react similarly to achiral molecules. This is because their lack of optical activity prevents them from exhibiting different interactions with chiral environments or reagents. This aspect is essential in various chemical and biochemical processes, where chiral recognition plays a significant role. For instance, in enzymatic reactions, which are often highly stereospecific, a meso compound might behave distinctly from its chiral counterparts, often exhibiting a lack of selective enzymatic action. This is a vital point to remember when working with reactions involving biological systems or using chiral catalysts.

The identification and characterization of meso compounds are often accomplished using a combination of techniques. Polarimetry, the measurement of optical rotation, directly reveals the lack of optical activity characteristic of these compounds. Furthermore, sophisticated spectroscopic methods like nuclear magnetic resonance (NMR) spectroscopy provide valuable insights into the molecular structure and symmetry, confirming the presence of a plane of symmetry. Advanced techniques like X-ray crystallography offer a definitive method of determining the exact three-dimensional structure of a molecule and confirming its symmetry properties. Understanding how to interpret NMR spectra or X-ray diffraction data is vital for the rigorous identification of meso compounds.

To summarize, meso compounds are unique stereoisomers exhibiting internal symmetry despite the presence of chiral centers. Their achiral nature significantly influences their reactivity and interactions with other chiral molecules. The ability to identify and characterize these compounds is a crucial skill for any chemist or biochemist. The understanding of meso compounds is fundamental to understanding optical activity and chirality and serves as a cornerstone for more advanced concepts in stereochemistry. Therefore, mastering the concept of meso compounds opens up numerous paths to further explore the multifaceted world of isomers and molecular variations. It's a fundamental concept for those hoping to master the intricacies of the fascinating field of isomerism.

5 Conformational Isomerism

Imagine a flexible molecule, like a beaded necklace you can twist and turn. That's essentially what conformational isomerism is all about. We're not changing the connectivity of the atoms—the beads stay strung together in the same order—but their spatial arrangement changes. This is different from the structural or stereoisomers we discussed earlier; here, the difference lies solely in the rotation around single bonds. Think of it as different poses the same molecule can adopt. These different poses are called conformers or rotamers. They can readily interconvert at room temperature, unlike the more stable isomers we've previously explored. This interconversion happens because rotation around single bonds is relatively easy.

Let's start with alkanes, the simplest hydrocarbons. Take ethane, for example, C2H6. Its two methyl groups (CH3) can rotate freely around the carbon-carbon single bond. However, not all rotations are equally favorable. In the staggered conformation, the hydrogen atoms on one methyl group are as far away as possible from the hydrogens on the other methyl group. This is the most stable conformation because it minimizes steric hindrance, the repulsion between electron clouds. The eclipsed conformation, on the other hand, puts the hydrogen atoms directly opposite each other, leading to increased repulsion and making it a higher-energy, less stable conformation. You can visualize this like the hands of a clock—the staggered conformation is like the hands at 6 o'clock, while the eclipsed conformation is like the hands at 12 o'clock.

Moving on to slightly more complex molecules, consider butane, C4H10. Butane has two methyl groups that can rotate around the central carbon-carbon bond. This rotation leads to several conformations, including anti, gauche, and fully eclipsed. The anti conformation, with the methyl groups 180 degrees apart, is the most stable due to minimal steric interactions. The gauche conformations, with the methyl groups 60 degrees apart, are less stable because of steric crowding. The fully eclipsed conformation is the least stable due to the maximum steric hindrance between the methyl groups. Understanding these different energy levels is crucial in predicting the molecule's behavior and reactivity. Imagine trying to build a model of these molecules—you'd find it easier to build the more stable conformations because they're less "crowded."

Now let's tackle ring conformations. Cycloalkanes, unlike the open-chain alkanes we've already discussed, present a different challenge. They have a fixed ring structure, which restricts the free rotation of their bonds. This restriction dramatically influences the overall shape and stability of the molecule. Let's focus on cyclohexane, a six-membered ring. The most stable conformation for cyclohexane is the chair conformation, where the ring adopts a three-dimensional structure that minimizes steric interactions. In this conformation, the six carbon atoms are not planar but are arranged in a chair-like shape. This conformation has both axial and equatorial positions for substituents. Axial substituents are perpendicular to the plane of the ring, while equatorial substituents are almost parallel to the plane.

The alternative conformation for cyclohexane is the boat conformation. In the boat conformation, the molecule is less stable than the chair conformation due to significant steric interactions. This is partly due to the presence of flagpole hydrogens, which are close to each other and experience steric repulsion. This repulsion increases the overall energy of the molecule. The boat conformation is, therefore, a higher-energy, less stable conformer than the chair conformation. It's crucial to understand that the chair and boat conformations are in equilibrium, but the chair conformation predominates due to its higher stability. You can think of it as a rocking chair \item it's constantly shifting between conformations but spends most of its time in the comfortable chair conformation. The energy difference between these conformations influences their relative populations.

Larger ring systems present even more complex conformational analyses. For instance, cyclopentane's preferred conformation is an envelope shape to reduce steric hindrance. Considering these different conformations and understanding their relative stabilities is crucial in various fields. For instance, in drug design, the specific conformation a molecule adopts can significantly influence its interactions with receptors, affecting its effectiveness. The spatial arrangement of atoms is critical for biological activity, and conformational changes can either enhance or decrease the drug's efficacy. This understanding of conformational isomerism helps chemists and pharmaceutical scientists design more efficient and effective drugs. Essentially, they need to know which conformations are most favorable for the desired biological activity. This is where computational chemistry often steps in to predict and model these various conformations.

Conformational isomerism is a key concept in understanding molecular structure and behavior. It illustrates how the three-dimensional arrangement of atoms within a molecule, even without changes in connectivity, can significantly influence its properties and reactivity. Understanding the different conformations a molecule can adopt, as well as their relative stabilities, is vital for predicting its behavior in chemical reactions and its interactions with other molecules. This is of particular importance in fields like medicinal chemistry, materials science and polymer chemistry, where the shape of a molecule heavily dictates its functionality.

5.1 Conformations of Alkanes

Alkanes, the simplest hydrocarbons, might seem unremarkable, but even these molecules exhibit fascinating isomerism in the form of conformations. These aren't distinct molecules like structural or stereoisomers; instead, they represent different spatial arrangements of the same molecule, arising from the free rotation around carbon-carbon single bonds. Think of it like twisting a flexible straw—you're changing its shape, but not its fundamental composition.

The most basic example is ethane, CH. Its two carbon atoms are connected by a single bond, allowing rotation around this bond. Two key conformations emerge: staggered and eclipsed. In the staggered conformation, the hydrogen atoms on one carbon are as far apart as possible from those on the other carbon, minimizing repulsive forces. This is the more stable conformation, representing a lower energy state. The eclipsed conformation, conversely, places hydrogen atoms directly opposite each other, leading to greater steric hindrance and higher energy. The energy difference between these conformations, though, is relatively small, meaning that they readily interconvert at room temperature. This constant interconversion is a key characteristic of conformations. You'll never isolate just one conformation of ethane; instead, you'll have a dynamic equilibrium between the staggered and eclipsed forms.

Moving to larger alkanes like propane (CH) and butane (CH), the conformational complexity increases significantly. In propane, the rotation around the central C-C bond generates various staggered and eclipsed conformations, again with staggered being energetically favored. Butane introduces an additional layer of complexity because it has a central C-C bond flanked by two methyl groups. Now, we have to consider the interactions between these methyl groups, adding anti, gauche, and totally eclipsed conformations to the mix. The anti conformation, where the methyl groups are furthest apart (180° dihedral angle), is the most stable. Gauche conformations, with methyl groups closer (approximately 60° dihedral angle), are less stable due to steric clashes, but they are still quite significant in the equilibrium.

Visualizing these conformations using Newman projections is incredibly helpful. This simple representation allows you to easily see the relative positions of the atoms and appreciate the steric interactions that govern conformational stability. Drawing these projections will help you understand the differences between staggered and eclipsed forms, and the energy differences between various conformations of butane, such as the anti and gauche forms. Practice is key; the more you draw, the better you'll grasp the concept. Learning how to represent the molecule differently allows you to solve different types of isomerism problems.

The impact of steric interactions extends beyond simple alkanes. In more complex molecules, the presence of bulky substituents can significantly influence the preferred conformations. This influence can profoundly impact a molecule's properties, reactivity, and overall behavior. Think about

how this affects proteins, for example; the unique 3D shape of a protein is a direct result of the conformations adopted by its amino acid chain. This intricate three-dimensional structure is vital for its function. These conformations aren't static; they constantly shift and rearrange, a dynamic interplay influenced by factors such as temperature and solvent.
The study of alkane conformations might seem like an academic exercise, but its implications are far-reaching. Understanding the relative stabilities of different conformations is vital for predicting molecular properties and reactivity. This knowledge is critical in numerous fields, including drug design, materials science, and polymer chemistry, where the shape of molecules directly impacts their functionality. By understanding these fundamental principles, you'll be better equipped to approach more complex problems involving isomerism. This chapter is designed to serve as a stepping stone to a deeper comprehension of the intricacies of isomerism, allowing you to approach the world of molecular structures with a newfound appreciation for its fascinating and versatile nature. Ultimately, mastering the concepts outlined here forms a crucial foundation for more advanced studies in organic chemistry and related disciplines. Think of this as your gateway to understanding the elegant complexity inherent within even the simplest of molecules. The implications of these simple yet profound molecular variations extend far beyond the classroom.

5.2 Ring Conformations

Cyclohexane, a seemingly simple molecule, offers a fascinating introduction to ring conformations. Its six-membered ring doesn't lie flat like a hexagon drawn on paper; instead, it adopts three-dimensional shapes to minimize steric strain – the repulsion between atoms that are too close together. Imagine trying to force six people to sit equally spaced around a tiny circular table – they'd be bumping elbows! Cyclohexane cleverly avoids this uncomfortable situation by twisting its structure.

The most stable conformation is the chair conformation. Picture a chair with its backrest and seat – that's essentially the shape of the cyclohexane ring. In this arrangement, the carbon atoms are not all in the same plane. Some carbons point upward (axial positions), while others point downward (equatorial positions). This staggered arrangement keeps the hydrogen atoms well apart, leading to reduced steric strain and increased stability. Think of it as the most comfortable seating arrangement for our six people.

Equatorial positions are generally preferred over axial positions due to less steric hindrance. Axial substituents experience 1,3-diaxial interactions, meaning they're uncomfortably close to hydrogens on carbons three positions away. Equatorial substituents, on the other hand, enjoy more space. The difference in energy between axial and equatorial positions is relatively small, however, often making isomerization between chair conformations feasible. This means the molecule can easily flip between two chair conformations, interconverting its axial and equatorial substituents.

The less stable conformation is the boat conformation. In this form, the ring resembles a boat, with two "flagpoles" at opposite ends. This conformation suffers from considerable steric strain because several atoms are forced into close proximity, causing significant repulsions between them. Imagine our six people at the table now crammed together in a small boat – not very comfortable, and certainly less stable than the spacious chair arrangement.

Beyond the chair and boat conformations, other less stable conformations exist, such as the twist-boat conformation, which represents a transition state between boat and chair forms. It's like a boat slightly distorted, still less stable than the chair form due to the remaining steric hindrance. The energy barriers separating these different conformations are relatively low, allowing for rapid interconversion at room temperature.

Understanding ring conformations is crucial in organic chemistry because they affect the reactivity and properties of molecules. For example, the reactivity of a substituent on a cyclohexane ring depends on whether it's in an axial or equatorial position. A bulky substituent in an axial position will experience significant steric hindrance, making it less reactive than the same substituent in an equatorial position. This is a direct consequence of the spatial arrangement dictated by the ring conformation. Understanding these subtle spatial differences helps predict chemical behavior.

Furthermore, the principles of ring conformations extend beyond cyclohexane. Other cyclic molecules, like cyclopentane and larger rings, also adopt specific conformations to minimize energy. However, the complexity increases with ring size. Larger rings often exhibit greater flexibility, leading to a wider range of possible conformations. While cyclohexane is a convenient starting point, it serves as a foundational example for understanding how ring structures arrange themselves in three-dimensional space.

Substituted cyclohexanes provide even more intriguing examples. The preferred conformation of a substituted cyclohexane ring is the one that places the largest substituents in equatorial positions, minimizing steric strain. If you have a cyclohexane ring with two different substituents, you might encounter different possibilities, leading to the formation of stereoisomers, specifically cis and trans isomers, depending on the relative positions of the substituents. Cis isomers have substituents on the same side of the ring, while trans isomers have substituents on opposite sides. These subtle differences can greatly influence molecular properties and reactivity.

Finally, remember that the concepts discussed here are essential building blocks for further explorations in organic chemistry. The ability to visualize and understand the three-dimensional structure of molecules is crucial for comprehending their properties and predicting their behavior. As you progress in your studies, you'll encounter more complex ring systems and develop more sophisticated techniques for predicting their conformations and understanding their reactivity based on the positions of substituents. This foundational understanding will allow you to analyze more complex scenarios and molecules confidently.

6 Applications of Isomerism

The impact of isomerism extends far beyond the theoretical realm; it's a cornerstone of many fields. Consider pharmaceuticals, for instance. The subtle difference between two isomers can dramatically alter a drug's efficacy and even its toxicity. Think of thalidomide, a tragic example where one isomer provided relief from morning sickness while the other caused severe birth defects. This underscores the critical importance of isomeric purity in drug development and underscores the need for rigorous testing before a drug can be deemed safe for human use. Understanding isomerism is not just an academic exercise; it's a matter of life and death. This profound impact highlights the deep importance of isomeric purity and the sophisticated analytical techniques required to determine the exact composition of drug molecules. Each step in the drug manufacturing process, from synthesis to formulation and distribution, necessitates careful consideration of isomeric purity and its potential consequences. The difference between isomers may be small at a molecular level, but the effects can be catastrophic without the proper attention.

Nature itself utilizes isomerism as a powerful tool. Many naturally occurring compounds, from sugars to amino acids, exist as isomers. The specific isomer present often determines the molecule's biological function and its interaction within a complex biological system. For example, the different isomers of glucose exhibit varying degrees of metabolic activity within our cells, illustrating nature's clever use of isomers to fine-tune biochemical processes. Similarly, the specific configuration of amino acids plays a critical role in protein folding and enzyme function, again showing how the intricate interplay of isomers is crucial for life. Investigating the role of isomerism in such biological systems provides immense insights into the complexity of living organisms and paves the way for developing new pharmaceuticals and therapeutic strategies.

The influence of isomerism is far-reaching, significantly impacting various industrial processes. The production of polymers, for instance, is heavily dependent on the isomeric composition of the starting monomers. Slight changes in the isomeric makeup of these monomers lead to substantial alterations in the final polymer's properties. The resulting polymer's mechanical strength, flexibility, and other characteristics can change dramatically depending on the relative abundance of different isomers involved in the polymerization process. This allows for a wide spectrum of polymer applications, from flexible plastics to strong structural materials. Similarly, the refining and processing of petroleum products heavily rely on isomerization processes to convert less desirable hydrocarbon isomers into those that are more valuable and widely usable. These industrial applications highlight the practical and economical importance of understanding and controlling isomerism in various chemical processes. The ability to selectively produce and manipulate specific isomers often leads to significant advancements in industry. The control over isomerism can optimize material properties and enhance the overall efficiency of many manufacturing processes.

Beyond its relevance in pharmaceuticals, biological systems, and industrial processes, isomerism presents a remarkable area of research. Current research focuses on developing more efficient methods for isomer separation and identification. This involves applying sophisticated techniques such as chromatography, spectroscopy and X-ray crystallography. Advanced computational techniques are also aiding in the prediction and analysis of isomeric properties, allowing for greater understanding and manipulation of these molecules. The continued study of isomerism extends to explorations into novel applications, from materials science to advanced drug development, signifying the enduring significance of this field in contemporary and future scientific and technological advancements. Exploring new techniques for isomer separation is critical to improving efficiency and reducing waste within industrial processes.

The field of isomerism continuously evolves, with ongoing research paving the way for novel applications and a deeper understanding of the molecular world. From developing new pharmaceuticals to refining industrial processes, isomerism is integral to many aspects of our lives. Its profound impact should not be underestimated. The continued exploration and understanding of isomerism holds the potential to lead to significant scientific breakthroughs and technological innovations in the future. This makes the study of isomers and isomerism an exciting and ever-evolving field within chemistry. The practical implications of this fundamental aspect of molecular structure are wide-ranging and hold significant promise for future advancements in various scientific and technological fields. The understanding and application of isomerism are constantly evolving, offering endless opportunities for innovation and scientific advancement.

6.1 Isomerism in Pharmaceuticals

The pharmaceutical industry is deeply intertwined with the fascinating world of isomerism. Consider ibuprofen, a common pain reliever. Its effectiveness relies entirely on one specific isomer, (S)-ibuprofen; the other isomer, (R)-ibuprofen, is largely inactive. This stark difference highlights the critical role isomeric purity plays in drug development and efficacy. Understanding this, therefore, is crucial for anyone aspiring to a career in drug design or related fields. Think of the implications—developing drugs without considering isomeric forms can lead to ineffective treatments, wasted resources, and potentially, harmful side effects.

Manufacturing drugs with a single isomer, instead of a mixture, is often preferred. Producing pure enantiomers can be challenging and significantly increases production costs. However, the benefits often outweigh the expense. For instance, thalidomide, a drug infamous for its devastating side effects, tragically illustrates this point. It was sold as a racemic mixture, but only one isomer possessed the desired sedative properties, while the other caused severe birth defects. This harrowing example firmly established the critical need for isomeric purity in pharmaceutical development and highlights the long-term consequences of ignoring this crucial aspect of drug design.

The impact of isomerism extends beyond just efficacy; it also impacts how a drug is metabolized and excreted by the body. Different isomers can interact with enzymes and receptors in distinct ways, leading to varying rates of absorption, distribution, metabolism, and excretion (ADME). These differences can dramatically affect the drug's duration of action and its overall impact on the body. For example, some isomers may be metabolized more quickly than others, requiring more frequent dosing, while others may persist in the body for extended periods, potentially increasing the risk of adverse effects.

Understanding the nuances of isomerism requires delving into various techniques used to identify and separate different isomers. Techniques like chiral chromatography, polarimetry, and nuclear magnetic resonance (NMR) spectroscopy are indispensable tools for characterizing isomeric mixtures and isolating specific isomers. These methods are crucial in ensuring the quality and safety of pharmaceutical products. Sophisticated spectroscopic analysis can reveal the subtle structural differences that distinguish isomers, enabling precise identification and quantification of each component in a mixture.

Beyond the practical aspects, studying isomerism in pharmaceuticals fosters an appreciation for the intricate relationship between molecular structure and biological activity. It encourages a deeper understanding of how even minuscule changes in a molecule's three-dimensional structure can profoundly alter its interaction with biological systems. This knowledge opens up exciting possibilities for drug design, allowing scientists to fine-tune molecular structures to optimize efficacy while minimizing potential side effects. By understanding the subtleties of how specific isomers bind to specific receptors and enzymes, we can create more targeted therapies and improve human health.

This chapter has touched upon the crucial role isomerism plays in the pharmaceutical field. However, further exploration into specific examples of drug isomers and their differing pharmacological properties could be illuminating. The specific impact of different isomeric forms on specific diseases or pathways would also be worth further investigating. Looking ahead, the ever-evolving field of medicinal chemistry promises continued innovations in the area of isomer-specific drug design, refining our understanding of these crucial molecules and improving the lives of countless people around the world. It is a complex yet fascinating area of study, full of both practical applications and fundamental scientific concepts. By grasping these foundational aspects, you can better appreciate the true elegance and complexity of this science, ready to face the challenges of tomorrow. The field of pharmacology is not static; ongoing research continuously refines our understanding, leading to the development of ever more sophisticated and effective drugs. This ongoing process of discovery and refinement underlines the enduring importance of understanding isomerism in the context of medicinal chemistry. The future holds exciting prospects for discovering and understanding the intricate connections between molecular structure and biological activity, ultimately leading to new breakthroughs in drug development.

6.2 Isomerism in Nature

Nature, a master chemist, orchestrates a breathtaking display of isomerism. From the vibrant hues of a butterfly's wings to the complex scents of a blooming flower, isomeric variations play a crucial, often unseen, role. Consider the multitude of terpenes, volatile organic compounds found in essential oils derived from plants. These molecules, often featuring a backbone of isoprene units, exist as a dazzling array of isomers, each contributing uniquely to a plant's characteristic aroma and biological activity. Limonene, for example, exists as two enantiomers: one smells like lemons, the other like oranges – a striking difference originating solely from a subtle spatial arrangement of atoms. This illustrates the profound impact of even minor structural alterations on the perception and function of natural compounds.

Think about the sugars we consume daily. Glucose, fructose, and galactose share the same molecular formula, CHO, yet their distinct structures determine their different metabolic pathways and sweetness levels. Glucose, a primary energy source, has a linear chain and a specific cyclic form. Fructose, a ketohexose, possesses a different cyclic form and a sweeter taste. Galactose, a component of lactose, has yet another unique structure, crucial for its role in various biological processes. These subtle differences in isomeric form are essential for the functioning of our bodies.

The world of proteins provides even more complex examples. Proteins are chains of amino acids, each amino acid a chiral molecule possessing a chiral carbon atom. The specific arrangement of these amino acids dictates a protein's three-dimensional structure, a crucial factor determining its function. A slight change in the sequence, thus creating different isomers, can have a drastic effect – leading to a protein with a different shape and, therefore, a different biological role or even causing a disease. For instance, the substitution of a single amino acid in the hemoglobin molecule can lead to sickle cell anemia.

Similarly, consider the diversity of carbohydrates found in nature – starch, cellulose, and glycogen. All are polymers of glucose, yet their differing linkages and overall structural arrangements impact their properties dramatically. Starch, used for energy storage in plants, is easily digestible, while cellulose, the main structural component of plant cell walls, is indigestible by humans. This difference stems directly from the different arrangements of glucose units in the polymeric chains—a beautiful exam-

ple of isomerism defining macroscopic properties.
Even the nucleic acids, DNA and RNA, which carry genetic information, rely on isomeric variations. The sugar component in DNA, deoxyribose, is a slightly different isomer than the ribose found in RNA, affecting the stability and functionality of each molecule. This subtle difference plays a vital role in how genetic information is stored and expressed, highlighting the fundamental role isomerism plays in the building blocks of life. Understanding isomerism is not just about memorizing names and structures; it is about appreciating the intricate way nature crafts functional molecules from the same basic building blocks. The subtle differences between isomers drive the extraordinary diversity and complexity of the natural world, affecting everything from the aroma of a flower to the function of a protein. The consequences are enormous, shaping the very fabric of life as we know it. The next time you admire the beauty of nature, remember the hidden role of isomerism, subtly shaping the world around us.

6.3 Industrial Applications

The impact of isomerism extends far beyond the laboratory, significantly influencing various industrial processes. Consider the production of polymers, the backbone of countless everyday materials. The precise arrangement of monomers, whether it's head-to-tail or head-to-head, dramatically alters the resulting polymer's properties – its strength, flexibility, melting point, and even its reactivity. This control over isomeric forms allows engineers to tailor polymers for specific applications, from flexible plastics to robust fibers. Understanding this allows for the creation of materials with specific properties, and the design of more efficient production methods to minimize waste and optimize yields.

Petrochemical industries rely heavily on isomerization processes. Crude oil, a complex mixture of hydrocarbons, contains numerous isomers. Refineries employ catalytic isomerization to convert less desirable isomers into more valuable ones, optimizing fuel efficiency and improving the overall quality of gasoline and other petroleum products. For example, linear alkanes can be converted into branched isomers, leading to a higher octane rating and reducing knocking in engines. The specific catalysts used and reaction conditions are meticulously designed to favour the formation of the desired isomer. This fine-tuned control showcases the practical importance of isomeric selectivity in large-scale chemical operations.

The food industry also interacts with isomerism. Many food additives and flavorings are specific isomers, with different isomers exhibiting vastly different tastes and aromas. For instance, the isomeric forms of carvone, a terpene, are responsible for the distinct flavors of spearmint and caraway. Manufacturers precisely control the production of the desired isomer to achieve the targeted taste or smell. This detailed attention to isomeric purity is particularly important in the production of artificial sweeteners and flavor enhancers, where even subtle changes in structure can significantly affect the perceived taste and palatability of the final product. This precise control highlights the sophisticated level of isomeric manipulation applied in the food industry.

Beyond food and fuel, many industrial catalysts are designed to work only with specific isomers. This selectivity is not merely a consequence of accident or limitations. Rather, it is deliberately engineered into the catalyst, by controlling its composition and design, to ensure efficiency and effectiveness. This controlled interaction is crucial in streamlining chemical reactions and reducing undesired side products. Understanding this selectivity is paramount in optimizing catalyst design and application. This results in cost savings and an environmentally friendly approach to chemical synthesis. Therefore, comprehending and controlling isomerism directly impacts the efficiency and sustainability of many industrial processes.

The production of pharmaceuticals relies heavily on the ability to synthesize specific isomers. Many drugs exist as enantiomers, and often only one enantiomer possesses the desired therapeutic effect. The other enantiomer might be inactive or even harmful. The efficient production of a single, desired enantiomer requires sophisticated techniques and a deep understanding of stereochemistry and chiral chemistry. This is crucial for safety and efficacy in medicine. The development and manufacture of these medications represent a direct industrial application of our advanced understanding of isomerism, significantly impacting public health and well-being. Indeed, the careful control of isomeric purity during pharmaceutical production is a non-negotiable aspect of drug manufacturing processes.

In conclusion, the influence of isomerism on various industries is undeniable, from the modification of polymer properties to the precise synthesis of pharmaceuticals. Understanding isomerism and the ability to control it allows for the development of more efficient, sustainable and safer industrial processes. The ability to selectively produce specific isomers has far-reaching implications for the chemical, food, and pharmaceutical industries, amongst many others. It represents a cornerstone of modern chemical engineering and manufacturing. It's a field ripe for further innovation and exploration, promising even more impactful advancements in the future. The impact of isomerism remains a vital aspect of the ongoing pursuit of more efficient, economical, and environmentally responsible industrial practices. Therefore, the mastery of isomeric control will continue to drive innovation across numerous sectors.

7 Advanced Isomerism Concepts

Analyzing complex isomers can be challenging, especially when dealing with molecules possessing multiple chiral centers or intricate ring systems. For instance, consider a molecule with four chiral carbons; the number of possible stereoisomers explodes, demanding sophisticated techniques for separation and identification. High-performance liquid chromatography (HPLC), coupled with chiral stationary phases, allows for the separation of enantiomers and diastereomers based on their subtle differences in interactions with the column material. Nuclear magnetic resonance (NMR) spectroscopy, particularly advanced techniques like 2D NMR, provides invaluable structural information. Analyzing coupling constants and chemical shifts allows for the precise determination of the relative configurations of chiral centers within the molecule. Mass spectrometry (MS) offers complementary information, determining the molecular weight and fragmentation patterns which can help identify different isomers. In essence, a combination of these techniques often proves necessary to fully elucidate the structure of complex isomers. This multifaceted approach is critical for researchers in various fields, including medicinal chemistry where the precise identification of isomers is essential for determining biological activity and safety profiles.

Modern techniques significantly enhance our ability to study isomerism. For example, X-ray crystallography provides detailed three-dimensional structural information at the atomic level, resolving even subtle conformational differences. This technique offers definitive proof of a molecule's structure. However, crystallization is not always easy, so computational chemistry has emerged as a powerful tool. Molecular modeling and simulation techniques allow researchers to predict and analyze the properties of isomers, even those which are difficult or impossible to synthesize or isolate experimentally. These methods provide insights into molecular geometry, stability, reactivity, and various spectroscopic properties, facilitating the identification of isomers. Furthermore, advanced computational methods, such as density functional theory (DFT), provide highly accurate predictions of molecular properties, offering another layer of confirmation to experimental findings. This synergy between experimental and computational approaches is increasingly important for the advancement of isomerism research.

Future research in isomerism holds tremendous potential. The development of new chiral separation techniques, capable of resolving complex mixtures with higher efficiency and selectivity, remains a priority. This includes the design of novel chiral stationary phases and improved chromatographic methods. Understanding the influence of subtle structural variations on the properties of materials is also key. For example, exploring the impact of different isomers on the physical and chemical properties of polymers and other macromolecules promises to unlock new possibilities in materials science. Similarly, advancements in the study of biological systems will shed more light on the roles of different isomers in biological processes, particularly in relation to drug discovery and the development of targeted therapies. The field of dynamic stereochemistry, which explores isomer interconversions, will provide a deeper understanding of reaction mechanisms and kinetics, leading to the development of more efficient and selective synthetic strategies. Consequently, continuous progress in both experimental and theoretical approaches will undoubtedly deepen our comprehension of the diverse world of isomers, opening up exciting possibilities in chemistry and beyond.

7.1 Analyzing Complex Isomers

Analyzing complex isomers often requires a multifaceted approach, combining various techniques and a deep understanding of fundamental isomeric principles. For example, consider a mixture containing several stereoisomers, each with potential variations in their functional groups and connectivity. Simple techniques like melting point determination might not suffice to differentiate between these subtle structural variations. You'll need more sophisticated methods.

NMR spectroscopy, especially with advanced pulse sequences like COSY (Correlation Spectroscopy) or NOESY (Nuclear Overhauser Effect Spectroscopy), becomes invaluable. COSY helps identify the connectivity of protons within the molecule, revealing the carbon backbone and the placement of functional groups. NOESY, on the other hand, provides crucial spatial information, mapping the proximity of protons and helping distinguish between different stereoisomers. You might see signals indicating a specific spatial relationship between atoms that only certain isomers would possess. Think of it like a molecular map, detailing not just the roads but also the distance between buildings.

Mass spectrometry provides complementary information. By analyzing the fragmentation patterns of the molecules, we can deduce their molecular weight and fragmentation pathways. Characteristic fragments help distinguish between isomers with different functional groups or chain lengths. For instance, an isomer with a particular functional group might yield a specific fragment ion with a unique mass-to-charge ratio, serving as a definitive fingerprint. This approach is particularly useful when dealing with a large number of possible isomers.

Infrared (IR) spectroscopy is another useful technique, although it's often less specific than NMR or mass spectrometry. Different functional groups absorb IR radiation at characteristic wavelengths, which allows for the identification of these groups. However, since many isomers possess the same functional groups, IR spectroscopy alone often isn't sufficient for complete isomeric analysis. It's more of a complementary technique, helping you to eliminate some possibilities.

X-ray crystallography, if you can obtain suitable crystals of your isomers, provides the most definitive structural information. This technique precisely maps the three-dimensional structure of the molecule, revealing all bond lengths, bond angles, and the exact spatial arrangement of atoms. This is the ultimate tool for complex isomer analysis, giving you the definitive answer, but its reliance on obtaining suitable crystals makes it less accessible compared to NMR and mass spectrometry.

Beyond these instrumental techniques, computational chemistry plays an increasingly vital role. Software programs use theoretical models to calculate the properties of molecules, enabling predictions of NMR chemical shifts, vibrational frequencies (for IR analysis), and other spectroscopic parameters. By comparing these calculated values with experimentally determined data, you can not only confirm the identity of specific isomers but also gain insights into their relative stabilities and reactivities. This becomes increasingly important as isomers become more complex and their properties harder to predict experimentally.

Remember, analyzing complex isomers is rarely a standalone procedure. It often involves an iterative process, combining data from several techniques to verify the identified isomers. You might initially use less expensive and easier methods like IR spectroscopy to narrow down possibilities before using more advanced techniques like X-ray crystallography or detailed NMR experiments. This strategy optimizes resources and ensures accurate structural determination. The key is careful interpretation and correlation of results obtained from all techniques. Sometimes, a seemingly minor discrepancy between the expected and observed properties can be the clue you need to unveil the true nature of the complex isomeric mixture you are working with. It's a puzzle that needs careful piecing together.

Ultimately, the successful analysis of complex isomers requires patience, attention to detail, and a solid understanding of the theory behind different spectroscopic and computational techniques. It requires using all your knowledge and skills in the most practical way.

7.2 Modern Techniques

Nuclear Magnetic Resonance (NMR) spectroscopy is a powerful technique for isomer identification. It works by exploiting the magnetic properties of atomic nuclei, specifically those with non-zero spin, like ^{1}H and ^{13}C. Different isomers, even subtle structural variations, exhibit distinct NMR spectral patterns due to variations in their chemical environments. The chemical shifts, coupling constants, and integration values provide detailed information about the connectivity and spatial arrangement of atoms, allowing for precise isomer differentiation. For example, ^{1}H NMR can distinguish between positional isomers by showing different chemical shifts for protons near functional groups in varying positions. Similarly, ^{13}C NMR provides information about the carbon skeleton and can help identify chain isomers. Advanced NMR techniques, such as two-dimensional NMR, further enhance this capability by revealing complex interactions between nuclei. For beginners, focusing on understanding basic peak interpretation is crucial before delving into the more intricate aspects of these advanced NMR experiments.

Mass spectrometry (MS) is another invaluable tool in the isomer identification arsenal. This technique measures the mass-to-charge ratio of ions, offering a fingerprint for each molecule. Isomers, while having the same molecular formula and thus the same molecular weight, often produce different fragmentation patterns upon ionization. This difference arises from structural variances influencing the bond strengths and fragmentation pathways. By analyzing the mass spectrum's characteristic peaks and their relative intensities, one can distinguish between isomers. For instance, chain isomers might exhibit variations in the abundance of fragments corresponding to the different lengths of alkyl chains. Similarly, the fragmentation patterns of positional and functional group isomers will often differ dramatically, providing distinct signals for each. The combination of MS with other techniques like chromatography greatly increases the analytical power, ensuring the proper identification of isomers.

Infrared (IR) spectroscopy helps analyze the functional groups present in a molecule. Different isomers, possessing varying functional groups or different spatial arrangements of the same functional groups, showcase distinctive IR absorption patterns. Each functional group exhibits characteristic vibrations at specific frequencies. These vibrations absorb infrared light, resulting in absorption bands in the IR spectrum. The presence, absence, or variations in the position and intensity of these absorption bands can be used to discriminate between isomers. For example, the presence or absence of a carbonyl group (C=O) stretching vibration clearly differentiates between ketones and alcohols with the same molecular formula. Similarly, the precise position of the C=C stretching vibration can provide insights into the stereochemistry around the double bond in geometric isomers. Remember though, IR alone is generally insufficient for complete isomer identification, however, it works well in conjunction with other spectroscopic methods.

X-ray crystallography provides a detailed three-dimensional structure of molecules. This highly powerful technique is particularly beneficial for confirming the structure of complex isomers. If the isomer can be crystallized, X-ray diffraction patterns yield data that, after complex mathematical analysis, provides the precise positions of atoms in the molecule. This provides definitive confirmation of the isomer's structure, including stereochemistry. For beginners, interpreting crystallographic data might seem daunting. However, the basic principle is straightforward: comparing the obtained atomic coordinates and bond lengths with those predicted for different isomers allows for unambiguous identification. While requiring specialized equipment and expertise, the results offer unparalleled accuracy, and it is truly the most powerful technique for isomer identification when applicable.

Computational chemistry provides a theoretical approach to studying isomer properties and predicting their behavior. Utilizing advanced software, various quantum chemical calculations can determine the energy, structure, and other properties of different isomers. The calculated energy differences often indicate the relative stability of different isomers, whereas predicted spectral properties, like NMR and IR signals, can be compared to experimental data for identification. Beginners can start with simpler methods such as molecular mechanics, which are less computationally demanding and yet still helpful in determining favorable conformations and relative energies. As expertise and computational resources grow, more sophisticated methods such as density functional theory (DFT) become accessible, providing more accurate results. Remember that the accuracy of computational methods depends on the computational parameters and the quality of the models used; thus, validation against experimental data remains essential.

7.3 Future Research

The field of isomerism, while seemingly well-trodden, still holds a wealth of undiscovered territory. Future research should prioritize the development of more sophisticated and efficient methods for separating and identifying isomers, particularly those with subtle structural differences. Current techniques, while powerful, often struggle with complex mixtures or isomers exhibiting similar physical properties. Imagine a world where we could rapidly and accurately analyze the isomeric composition of any sample, unlocking new possibilities in various fields. This necessitates exploring advanced separation techniques like microfluidic devices and novel chromatographic methods tailored for specific isomer types. Improved instrumentation and data analysis algorithms, possibly incorporating machine learning, will be crucial to achieving this goal.

We need a deeper understanding of the dynamic behavior of isomers. Many isomers interconvert readily, and this dynamic equilibrium impacts their reactivity and properties. Investigating the kinetics and thermodynamics of these interconversions, perhaps through advanced computational modeling and real-time spectroscopic techniques, promises to reveal fascinating insights into the factors governing isomeric stability and transformations. This improved understanding will not only enhance our predictive capabilities but also allow for the controlled manipulation of isomeric ratios, opening new avenues in synthesis and materials science.

The biological world is a treasure trove of isomeric complexity. Many biologically active molecules exist as isomers, and often, only one isomer exhibits the desired biological activity. Future research should focus on the precise mechanisms by which biological systems selectively recognize and utilize specific isomers, excluding others. Understanding these intricate processes could lead to groundbreaking advancements in drug design, biosensing, and biotechnology. Studying enzyme-catalyzed reactions involving isomers could provide crucial clues, guiding us toward the development of highly selective catalysts for industrial processes as well. This research demands a multidisciplinary approach, integrating organic chemistry, biochemistry, and structural biology.

Furthermore, there's a significant opportunity to explore the applications of isomerism in novel materials. The properties of materials, from strength to conductivity, are highly dependent on their molecular structure, including isomeric composition. By precisely controlling the isomeric ratios during material synthesis, we can fine-tune their properties to create advanced materials with tailored functionalities. This would require innovative synthetic strategies, potentially using environmentally friendly and sustainable approaches, alongside advanced characterization methods to fully understand the structure-property relationship. Imagine creating novel polymers with exceptional strength or developing highly efficient catalysts based on carefully selected isomers. The possibilities are immense.

Computational chemistry plays a vital role in advancing isomerism research. As computational power continues to increase, sophisticated simulations will allow us to study the properties and behavior of isomers with unprecedented accuracy. This includes predicting the stability of different isomers, simulating their reactions, and designing new synthetic routes. Combining these computational predictions with experimental results will further enhance our understanding of isomerism. The development of more accurate and efficient force fields and quantum mechanical methods tailored for isomeric systems is essential for unlocking the full potential of computational chemistry in this field.

The future of isomerism research is bright, promising breakthroughs in various scientific disciplines. By embracing advanced technologies, innovative approaches, and collaborative efforts, we can unlock a deeper understanding of this fundamental aspect of chemistry. The practical implications of such research will be widespread, revolutionizing fields ranging from medicine and materials science to environmental remediation and energy production. The development of more environmentally friendly and sustainable methods of isomer separation and synthesis is crucial, aligning the research with broader sustainability goals. This multidisciplinary approach is key to unraveling the intricacies of isomerism and harnessing its potential to address some of humanity's most pressing challenges. The impact of this research will be far-reaching, shaping future technologies and improving our quality of life. The potential benefits extend across numerous sectors, including healthcare, industry, and environmental science. Ultimately, the continued exploration of isomerism promises to yield remarkable advancements and open new frontiers in our understanding of the molecular world.

www.ingramcontent.com/pod-product-compliance
Ingram Content Group UK Ltd.
Pitfield, Milton Keynes, MK11 3LW, UK
UKHW020156290926
15471UKWH00005B/904

9 798349 408489